The Beginner's Guide to Data Science

Robert Ball • Brian Rague

The Beginner's Guide to Data Science

 Springer

Robert Ball
Weber State University
Ogden, UT, USA

Brian Rague
Weber State University
Ogden, UT, USA

ISBN 978-3-031-07867-5   ISBN 978-3-031-07865-1   (eBook)
https://doi.org/10.1007/978-3-031-07865-1

This Springer imprint is published by the registered company Springer Nature Switzerland AG
The registered company address is: Gewerbestrasse 11, 6330 Cham, Switzerland